BOOK SOLD
NO LONGER R.H.P.L
PROPERTY

Inside Animal Homes

Inside Beehives

Matthew Bates

PowerKiDS press

New York

Published in 2016 by The Rosen Publishing Group, Inc.
29 East 21st Street, New York, NY 10010

Copyright © 2016 by The Rosen Publishing Group, Inc.

All rights reserved. No part of this book may be reproduced in any form without permission in writing from the publisher, except by a reviewer.

First Edition

Editor: Sarah Machajewski
Book Design: Mickey Harmon

Photo Credits: Cover, p. 1 (bees) Diane Garcia/Shutterstock.com; cover, pp. 3–4, 6, 8, 10, 12, 14, 16, 18, 20, 22–24 (honeycomb texture) LilKar/Shutterstock.com; cover, pp. 3–4, 6, 8, 10, 12, 14, 16, 18, 20, 22–24 (magnifying glass); p. 5 Danita Delimont/Gallo Images/Getty Images; p. 7 PCHT/Shutterstock.com; pp. 9 (stinger inset), 17 Photo Researchers/Science Source/Getty Images; p. 9 (bee) jocic/Shutterstock.com; p. 9 (background) sl_photo/Shutterstock.com; p. 11 angelshot/Shutterstock.com; p. 13 Gherasim Rares/Shutterstock.com; p. 15 (inset) Anders Blomqvist/Lonely Planet Images/Getty Images; p. 15 (main) Harry Rogers/Science Source/Getty Images; p. 19 Horst Sollinger/Getty Images; p. 21 sumikophoto/Shutterstock.com; p. 22 (left bee) Protasov AN/Shutterstock.com; p. 22 (right bee) Peter Waters/Shutterstock.com.

Library of Congress Cataloging-in-Publication Data

Bates, Matthew.
 Inside beehives / Matthew Bates.
    pages cm. — (Inside animal homes)
 Includes bibliographical references and index.
 ISBN 978-1-4994-0873-7 (pbk.)
 ISBN 978-1-4994-0888-1 (6 pack)
 ISBN 978-1-4994-0920-8 (library binding)
 1. Bees—Juvenile literature. 2. Beehives—Juvenile literature. 3. Animals—Habitations—Juvenile literature. I. Title. II. Series: Inside animal homes.
 QL565.2.B38 2015
 595.79'9—dc23
                                    2015009242

Manufactured in the United States of America

CPSIA Compliance Information: Batch #WS15PK: For Further Information contact Rosen Publishing, New York, New York at 1-800-237-9932

Contents

At Home with Bees 4
Common Creatures. 6
Bee Body 8
Who's Inside the Hive? 10
Busy Worker Bees 12
Finding the Place 14
Preparing to Build 16
Building Combs 18
Surviving the Winter 20
Bees Are Important 22
Glossary 23
Index 24
Websites 24

At Home with Bees

What does the word "home" mean to you? For people, a home is a place to live, eat, sleep, and stay safe from the **conditions** outside. It's the same for many creatures in the wild, including bees.

Though bees don't live in homes that look like ours, the idea is the same. Their home is called a beehive. It's a safe place for them to live and grow. A bee's home may be one of the most interesting parts of this creature's life. Let's go inside to find out what it's like.

Bees depend on their home for **survival**.

Common Creatures

Bees are very common creatures. There are over 25,000 species, or kinds, in the world. They live everywhere except Antarctica. The bees you're probably most familiar with are honeybees and bumblebees.

Many kinds of bees live in large groups called colonies. There can be anywhere from 20,000 to 30,000 bees in one colony! They live together in a home called a hive. A hive may not look big enough from the outside, but bees know how to build a home that fits everyone.

Beehives can also be called nests. This nest holds thousands of bees.

Bee Body

Bees can be anywhere from 0.08 to 1.6 inches (0.2 cm to 4 cm) long. Their body is split into three parts—head, **thorax**, and **abdomen**. A bee has six legs, two antennae on its head, and two pairs of wings.

Each part of a bee's body has an important job. Tiny hairs covering its body help a bee collect pollen. Special mouthparts help it collect nectar. Bees bring whatever food they collect back to the hive.

THE INSIDE SCOOP

Bees drink nectar through a tube called a proboscis.

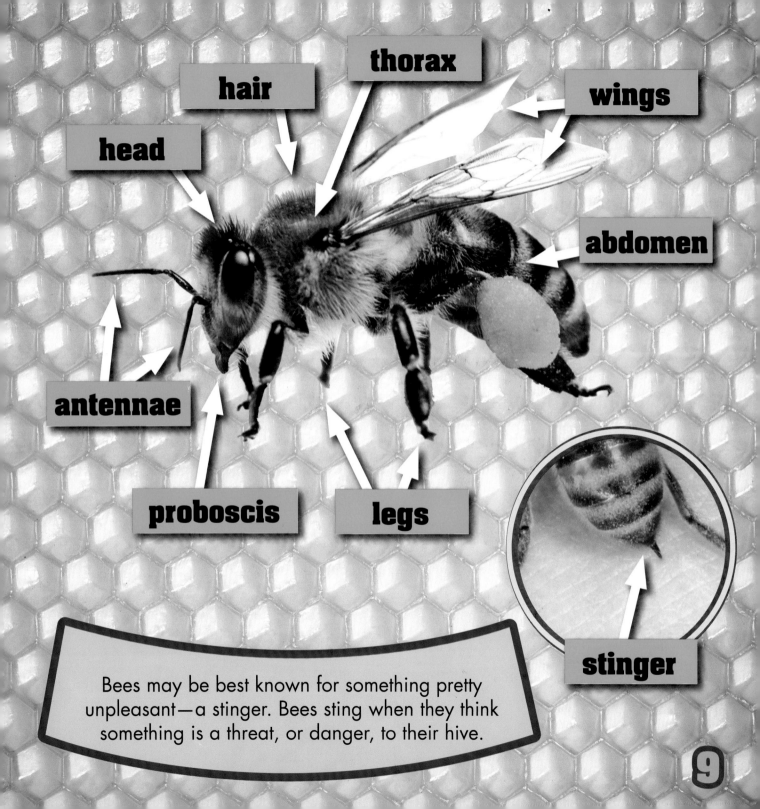

Bees may be best known for something pretty unpleasant—a stinger. Bees sting when they think something is a threat, or danger, to their hive.

Who's Inside the Hive?

Many kinds of bees are social bugs. Each bee in a colony has a job to do, and those jobs help the colony survive. The queen bee has the most important job. She's the largest bee in the hive and the only one that lays eggs.

Queens are picked by bees at birth. They feed one **larva** a special matter called royal jelly, which makes that bee grow big and able to lay eggs. Drones are male bees. Their only job is to **mate** with the queen, but they die after doing so.

Queen bees have a longer abdomen than other bees in a colony, which you can see in this picture.

Busy Worker Bees

The success of a colony depends on a third kind of bee—female worker bees. They definitely earn their name. Worker bees build hives and keep tiny rooms called cells clean. They care for the queen and her larvae. They leave the hive in search of food and bring it back for the others.

As worker bees collect nectar, they store it in their "honey stomach." Back at the hive, bees either eat the nectar right away or turn it into honey and store it for later.

THE INSIDE SCOOP

Worker bees collect pollen and carry it in pollen baskets, which are special areas on their hind legs. Pollen is used as food or can be stored in the hive.

These worker bees care for the colony's larvae, which are the white, sticky-looking objects in this photo. Nurse bees feed them honey and pollen until they grow up.

larvae

Finding the Place

The thousands of bees in a colony need somewhere to live. This is the job of the worker bees. They search for **structures** that seem like a good place for a hive. They choose cavities, or empty spaces inside objects.

Bees don't just choose any cavity, however. The cavity must be big enough to hold at least 6.5 gallons (24.6 l). The entrance can be no bigger than 1.5 inches (3.8 cm) wide. The cavity must also be high off the ground to stay safe from predators.

THE INSIDE SCOOP

Beekeepers are people who raise and care for bees. They build structures to give bees a place to live. The structures are often closed wooden boxes with narrow openings, much like the structures bees use in the wild.

In the wild, bees often build nests inside **hollow** trees, caves, and cracks in large rocks.

A bee sits in the entrance to its nest, which is a tiny tube leading to the inside of a tree.

Preparing to Build

Bees get to work once they've found the right place for a hive. They first scrape off anything on the walls of the entrance and cavity. For example, in a hollow tree, they will scrape the wood to create a smooth surface.

Next, the bees coat the inside of their future hive with a matter called propolis. Propolis is a sticky sap that comes from trees. Bees use it as a coating and to seal any open spaces in their hive.

THE INSIDE SCOOP

Propolis also helps keep diseases, or sicknesses, from entering the hive.

One of the first steps in building a hive is laying down propolis.

Building Combs

Then next stage in hive **construction** is making combs. Worker bees make wax inside their body. As the wax leaves their abdomen, bees shape it into combs with six-sided cells. Bees attach the combs to the roof and walls of the hive, building from the top down.

Each comb has two **layers** of six-sided, or hexagonal, cells. The cells are used for storing honey and pollen. Combs also contain cells that are used for raising bee larvae.

THE INSIDE SCOOP

A natural beehive has about 100,000 cells. It takes about 2.5 pounds (1.1 kg) of wax to make a hive this size.

queen cell

Cells are always hexagons. Using this shape is smart. Hexagons fit together with no spaces in between. This may mean bees use less wax than they would with other shapes.

Surviving the Winter

Bees are a common sight in summer, but in winter, bees stay in their hive. Unlike other creatures, bees don't hibernate, or go to sleep for a long period of time.

How do bees survive the winter? They eat their stores of honey and pollen. They keep the cold from entering the hive by sealing any holes in its walls. They huddle in one big group to keep warm. Bees also beat their wings to create heat. The inside of a hive must be at least 50 degrees Fahrenheit (10 degrees Celsius). Otherwise, the colony will die.

Bees spend all summer collecting nectar and pollen. They must store enough to feed the colony for an entire winter.

Bees Are Important

Bees' hard work helps their colony, but it helps people, too. Bees spread pollen between flowers as they fly around. This helps us grow the plants we use as food, which means we need bees to survive.

However, bees around the world are dying, possibly because of **chemicals** farmers spray on plants. Bees take the chemicals back to the hive, which hurts the whole colony. We can help bees and their hives by not treating our plants with harmful chemicals. It's one way to help these important creatures.

Glossary

abdomen: The part of a bug's body that holds the stomach.

chemical: Matter that can be mixed with other matter to cause changes.

condition: The state of something.

construction: The act of building.

hollow: Empty on the inside.

larva: A bug in an early life stage that has a wormlike form. The plural form is "larvae."

layer: One thickness lying over or under another.

mate: To come together to make babies.

structure: A building or other object.

survival: The state of continuing to live.

thorax: The part of a bug's body that holds the heart and lungs.

Index

A
abdomen, 8, 9, 11, 18
antennae, 8, 9

B
beekeepers, 14

C
cells, 12, 18, 19
chemicals, 22
colonies, 6, 10, 11, 12, 13, 14, 20, 21, 22
combs, 18

D
drones, 10

E
eggs, 10

H
head, 8, 9
honey, 12, 13, 18, 20

L
larvae, 10, 12, 13, 18
legs, 8, 9, 12

N
nectar, 8, 12, 21
nests, 6, 15

P
pollen, 8, 12, 13, 18, 20, 21, 22
proboscis, 8, 9
propolis, 16, 17

Q
queen bee, 10, 11, 12, 19

S
stinger, 9

T
thorax, 8, 9

W
wax, 18, 19
wings, 8, 9, 20
worker bees, 12, 13, 14, 18

Websites

Due to the changing nature of Internet links, PowerKids Press has developed an online list of websites related to the subject of this book. This site is updated regularly. Please use this link to access the list: www.powerkidslinks.com/home/bees